ARIZONA AGRICULTURE
BEE'S AMAZING ADVENTURE

BY BONNIE APPERSON JACOBS
AND TERRI MAINWARING

in cooperation with
The University of Arizona College of Agriculture and Life Sciences Cooperative Extension
and Maricopa County Farm Bureau

A division of Five Star Publications, Inc.
Great Books for Growing Minds
Chandler, Arizona

Text Copyright ©2014 The University of Arizona
Cover & Interior Design Copyright ©2014 Five Star Publications, Inc.

All Rights Reserved. No part of this book may be used or reproduced or transmitted in any form or by any means, electronic or mechanical, including photocopying, recording, or by any information storage or retrieval system without written permission except in the case of brief quotations used in critical articles and reviews. Requests for permissions should be addressed to the publisher:

Linda F. Radke, President
Five Star Publications, Inc.
PO Box 6698
Chandler, AZ 85246-6698
(480) 940-8250
www.FiveStarPublications.com

www.ArizonaAgricultureBook.com

A division of Five Star Publications, Inc.
Great Books for Growing Minds

Publisher's Cataloging-In-Publication Data

Jacobs, Bonnie Apperson.
   Arizona agriculture : Bee's amazing adventure / by Bonnie Apperson Jacobs and Terri Mainwaring ; in cooperation with The University of Arizona College of Agriculture and Life Sciences Cooperative Extension and Maricopa County Farm Bureau.
   pages : color illustrations ; cm
   Summary: From cotton bolls and citrus groves to cattle ranches and dairy farms, Pee Wee Bee takes an amazing journey through Arizona agriculture.
   Interest age level: 006-010.
   Issued also as an ebook.
   ISBN: 978-1-58985-267-9

   1. Bees--Juvenile literature. 2. Agriculture--Arizona--Juvenile literature. 3. Bees. 4. Agriculture--Arizona. I. Mainwaring, Terri. II. University of Arizona. Cooperative Extension. III. Maricopa County (Ariz.). Farm Bureau. IV. Title. V. Title: Bee's amazing adventure

S451.A6 J33 2014
630.9791

Five Star Publications, Inc. sets high standards to ensure forestry is practiced in an environmentally responsible, socially beneficial and economically viable manner.

2014933890

Electronic edition provided by

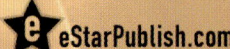

The eDivision of Five Star Publications, Inc.

Printed in the United States of America

Photographs: Tim Trumble Photography, Monica Kilcullen Pastor, Barbara Brady (Arizona Cattle Growers Association Photo Contest), Dave Massey (Shutterstock), Brent Murphree, Kurt Nolte (University of Arizona – Yuma), Linda Covey (Emily Brown, AZ Queen Bee), David W. Schafer (Arizona Cattle Growers Association Photo Contest), Arizona Farm Bureau Federation, The Biodesign Institute at Arizona State University, Goodluz (Shutterstock), Hickman's Family Farms, and Western Farm Press,
Supplemental Illustrations: Nadia Komorova
Cover Design & Page Layout: Kris Taft Miller
Development Editor: Jennifer Steele Christensen
Proofreader: Cristy Bertini
Project Manager: Patti Crane
Representatives for Cooperative Partners: Monica Kilcullen Pastor (UA CALS-CE) and Jeannette Fish (MCFB)

A portion of the funding for this project was provided by The Arizona Department of Agriculture, Agricultural Consultation and Training using Specialty Crop Block Grant funds from the USDA, Agricultural Marketing Service. The views or findings presented are the Grantee's and do not necessarily represent those of the Arizona Department of Agriculture, the state of Arizona, or the USDA.

"As we developed this project, we wanted students to understand how deeply their lives are impacted by agriculture. Farmers and ranchers play a critical role in our lives, and we are thrilled with this opportunity to help children appreciate how important agriculture is to us all."

— Jeannette Fish, executive director, MCFB and
Monica Kilcullen Pastor, associate programmatic agent, ag and natural resources, UA CALS-CE

DEDICATIONS

To my daughter, Megan, my favorite honey.
— Bonnie Apperson Jacobs

BUZ-Z-Z! This book is dedicated to my husband, Tom. His interest, patience, love, and sense of humor along the way has made this experience amaz-z-z-ing!
— Terri Mainwaring

ACKNOWLEDGEMENTS

The University of Arizona College of Agriculture and Life Sciences Cooperative Extension wishes to thank the following organizations for their contributions to the funding of this book:

The Arizona Department of Agriculture, Agricultural Consultation and Training has funded a portion of this book using Specialty Crop Block Grant funds provided by USDA, Agricultural Marketing Service. The views or findings presented are the Grantee's and do not necessarily represent those of the Arizona Department of Agriculture, the state of Arizona, or the USDA

Maricopa County Farm Bureau

Maricopa County Farm Bureau

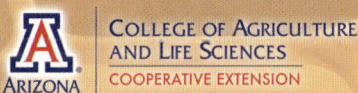

University of Arizona College of Agriculture and Life Sciences Cooperative Extension, Agricultural Literacy Program

UA Agri-Press Club Endowment

Arizona Cotton Growers Association

Thank you to Monica Kilcullen Pastor of The University of Arizona College of Agriculture and Life Sciences Cooperative Extension and Jeannette Fish of Maricopa County Farm Bureau for partnering with Five Star Publications, Inc. to give this book wings. Also, thanks to Tim Trumble of Tim Trumble Photography for lending your talents. Finally, special thanks to my Five Star team who always makes us shine, especially Patti Crane, Jennifer Christensen and Kris Taft Miller, for adding your special touches on each and every page.

— Linda F. Radke, Five Star Publications, Inc.

HI KIDS!

Pee Wee Bee here. I am one proud and gorgeous Arizona honeybee.

I want to invite you to join me for an **AMAZ-Z-Z-ING** adventure! We'll **Z-Z-Z-IP** across the state together, learning about Arizona agriculture.

We'll even meet up with some members of my bee family! Our role in Arizona agriculture is **very** important!

ARIZ-Z-Z-ONA.

You probably already know about our deserts, our hot weather, our cacti, and of course, our spectacular Grand Canyon. Arizona folks brag about these things ... I know I do!

But many of you don't know much about Arizona agriculture. Since I do, I want to help spread the word.

LET'S GET BUZ-Z-Z-ING ALONG!

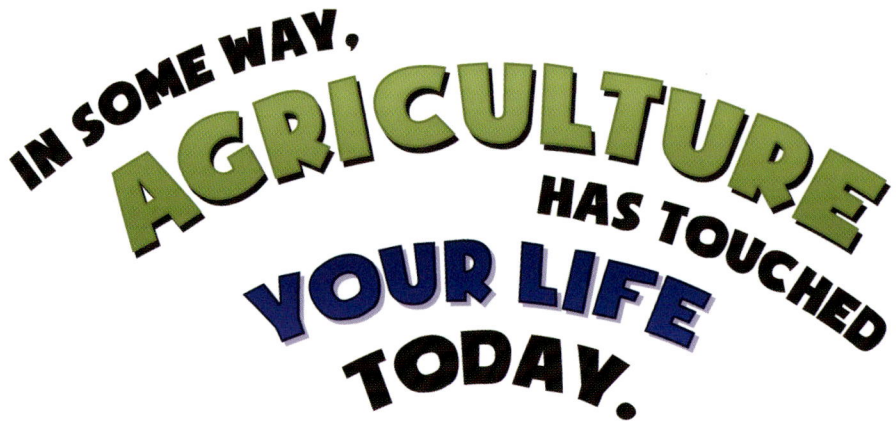

DID I HEAR SOMEONE ASK, "WHAT IS AGRICULTURE?" GREAT QUESTION!

Agriculture is all about growing **crops** and raising animals, using water carefully, and protecting the land.

Agriculture is an important business in Arizona, and it's how farmers and ranchers make a living. We have so much to learn together!

> MY UNCLE, BILL E. BEE, AND I SAY THAT FARMERS AND RANCHERS ARE AGRICULTURAL SCIENTISTS.

©Tim Trumble Photography

Farmers and ranchers play an important part in the lives of all Arizonans. Their hard work brings fruit, vegetables, eggs, milk, and meat to our dinner tables. Farmers and ranchers also produce cotton for clothes, grow hay to feed animals, and raise plants to bee-utify our homes.

©GoodLuz

I THINK GOING TO COLLEGE TO GET A DEGREE IN AGRICULTURE WOULD BEE FUN!

Farmers and ranchers go to college to study and learn about the best ways to farm and ranch. They use the latest and greatest technologies and machines to help solve problems, save time, work safely, and answer difficult questions.

Advancements in **technology** have made a big difference in the way we farm and ranch. How do you think technology improves Arizona's agriculture?

AUTO-STEER TRACTOR

Most of us have seen a tractor. But many of us don't know that some tractors can operate mostly hands-free. These are called auto-steer tractors and are controlled by computer GPS systems and other wireless technologies. These vehicles can plant seeds, fertilize plants, and dig the soil—and they won't run into people, animals, or objects.

©Tim Trumble Photography

MODULE BUILDING COTTON PICKER

A module building cotton picker allows one machine to do two jobs at once. This machine picks cotton from the plants and then packs it tightly together so it is easier to move.

©Tim Trumble Photography

SMARTPHONE APPLICATIONS

Many of today's farmers use "apps" on their smartphones to manage their water and keep track of the work they do on the farm.

©Tim Trumble Photography

RESEARCH

Farmers often talk to scientists to learn about new ways to grow crops, save water, improve their soil, and control pests.

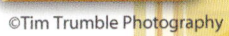
©Tim Trumble Photography

©Kurt Nolte, University of Arizona-Yuma

Arizona farmers grow most of the winter **lettuce** for America's salads. In fact, in the winter, Arizona is one of the only states with lettuce to sell. Because of our warm climate, lettuce is picked earlier here than anywhere else in the country.

As lettuce grows, its leaves open like a flower. The inner leaves form a solid ball or head. About 90 days after a lettuce seed is planted, a head of lettuce is ready to be picked and delivered to your grocery store.

LET'S DO LUNCH AT THE SALAD BAR! DELISH!

©Arizona Farm Bureau Federation

Along with lettuce, Arizona farmers grow many of the other ingredients we use to toss together a delicious salad. These are called **specialty crops** and include fruits, vegetables, and nuts. Other specialty crops include a wide variety of trees, plants, and flowers that are used in medicine and to bee-utify our homes.

WHAT'S THE BUZZ?

To harvest nuts, huge shaker machines jiggle each tree so the nuts drop into nets or onto the ground, ready for collection by special raker machines. Look out bee-low!

MY NIECE, BAR-BEE, AND I ARE SHAKIN' AND RAKIN'!

©Western Farm Press

©Monica Kilcullen Pastor

©Arizona Farm Bureau Federation

Pistachios and **pecans** are nuts grown on trees in southern Arizona. These trees produce a lot of nuts one year, and not as many the next year. This is called alternate bearing.

Melons are one type of specialty crop that Arizona farmers grow. Some of these crops include watermelons, cantaloupes, and honeydews, which can be planted and harvested twice a year; once in the fall and once in the spring.

"CHILE PEPPERS COME IN MANY DIFFERENT "HEATS," FROM MILD-TASTING TO "HOT-CHA-CHA!"

WHAT'S THE BUZZ?

Salsa has become even more popular in the USA than ketchup. It's "nacho" time!

Chile peppers are another important specialty crop in Arizona. These peppers are small, hot-tasting pods that come in different shapes, sizes, and colors. Salsa and other spicy sauces are made from chiles.

©Monica Kilcullen Pastor

EXCUSE ME WHILE I HELP MY COUSIN, MAY-BEE, POLLINATE A CITRUS TREE OR TWO!

Thanks to Arizona's warm **climate**, citrus can grow in several parts of the state. **Citrus** crops include oranges, lemons, grapefruit, limes, and tangerines. Citrus fruit begins growing in the springtime when the trees fill with blossoms. That's when bees **pollinate** the flowers.

When citrus fruit is growing, a farmer's two biggest worries are cold weather and insects. Both can harm the crop.

November through February are the very best months to pick citrus. Because the fruit is fragile, farmers have it picked by hand.

Farmers often invite families to pick their own fruits and vegetables straight from the orchards and fields. This is a fun way for families to work together and learn about Arizona agriculture.

WHAT'S THE BUZZ?

Because the trunks on citrus trees can get sunburned, they are sometimes painted white or tan. Paint on a tree trunk protects it from the sun, much like sunscreen.

©Monica Kilcullen Pastor

©Arizona Farm Bureau Federation

HEY, HEY, HEY!
DO YOU KNOW THE DIFFERENCE BETWEEN HAY AND STRAW?

©Tim Trumble Photography

Arizona farms produce more **hay** per **acre** than any other state in America.

In the winter, farmers plant seeds that keep growing for four months. As the hay gets taller, farmers mow it and leave it on the warm ground to dry. Then, they collect it with machines. Hay grows back and needs to be cut over and over, just like the grass on your school's playground!

Farmers keep cutting, baling, and storing hay all year long. One type of machine cuts the hay and another machine collects it and packs it into bales. A bale keeps the hay pressed together in a bundle that can be easily stored and transported.

WHAT'S THE BUZZ?

Hay is used for feed, and cows and horses love to eat it. **Straw** is used for bedding, and animals love to sleep on it.

©Tim Trumble Photography

Do farmers harvest cotton "bolls" or cotton "balls"? The words sound the same to me! Listen for the answer.

©Tim Trumble Photography

Arizona is one of the top ten cotton-producing states in America. Cotton's growing season begins in the spring. In the summertime, a boll forms after the cotton flower is pollinated. A boll is a hard shell on the plant that holds and protects the cottonseeds and fiber. In the fall, the boll pops open, displaying the fluffy white cotton inside. Then, it's harvest time!

A large machine called a cotton picker removes the cotton from the plant. Then, the harvested cotton is taken to a building where another machine called a cotton gin pulls out the seeds and forms the remaining cotton lint into bales, just like hay.

©Tim Trumble Photography

©Tim Trumble Photography

You already know that **cotton** is used to make fabric for clothing, but cotton is also valuable for so much more! Oil from cottonseeds is used in cooking, medicines, lamps, and lotions. It is also an ingredient in foods like crackers, salad dressing, and cereal. None of the plant is ever wasted in the cotton industry. There's an important use for it all.

©Monica Kilcullen Pastor

"I THOUGHT A NURSERY WAS A PLACE FOR BABIES!"

There's so much to love about Arizona agriculture! Green grass, colorful flowers, and leafy trees are also grown on Arizona farms and in greenhouses statewide. These products are sold in nurseries for families to buy. Why? To make our yards, homes, and businesses bee-utiful!

Some Arizona farms are **organic**. To grow crops organically, farmers use **fertilizers**, chemicals, and pest controls that are produced by nature. Some organic farmers release ladybugs or other helpful insects into their fields to eat the harmful bugs that damage crops. This is an example of **natural** pest control.

WHAT'S THE BUZZ?

Arizona families can purchase many kinds of organic foods at grocery stores or farmers' markets, including produce, meat, dairy products, and eggs.

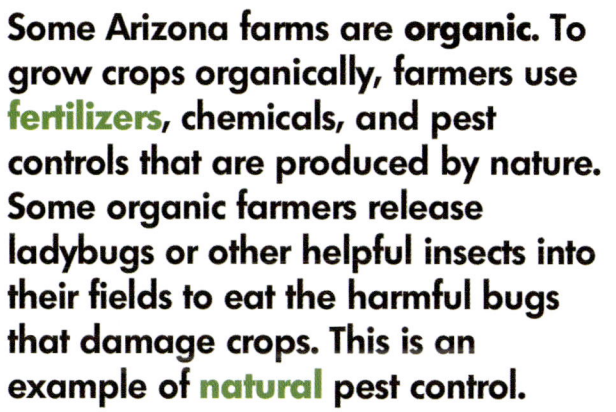

MY AUNT, HON E. BEE, POLLINATES ORGANIC MELONS AND STRAWBERRIES. SHE'S "BERRY" SWEET.

LET'S GET AN "EGGSPLANATION."

©Hickman's Family Farms

Arizona **egg** farmers worry about taking care of their hens, their farms, and their businesses. They don't worry about how the eggs are laid, because their hens know exactly what to do.

Hens live in henhouses, where water and nutritious grain are always available. Farmers rarely touch the eggs because modern machines do the day-to-day work.

Hens lay eggs that roll onto a moving **conveyor belt**. Machines then wash and size the eggs, check them for cracks, and pack them into cartons. Next, the eggs are carefully loaded onto trucks and delivered to stores. Most eggs reach Arizona grocery stores just one day after they are laid.

WHAT'S THE BUZZ?

One hen lays about five eggs per week. That's 260 eggs per year!

©Hickman's Family Farms

I LIKE MINE SCRAMBLED!

WHAT'S THE BUZZ?

Each day, a cow eats 100 pounds of food and drinks enough water to fill a bathtub.

If you like yogurt, cheese, ice cream, and milk, please thank a dairy farmer! Arizona **dairy** farmers take good care of their cows. That's why Arizona produces high-quality milk. Modern dairy farmers use special machines to milk their cows two or three times a day. Computers keep track of each cow's health and the amount of milk she produces.

Nutritionists recommend that people enjoy three cups of milk or milk products a day. Yogurt, cheese, ice cream, and cottage cheese are all dairy products.

Dairy farmers often raise their own hay to feed their cows. They save time and money when hay is grown on their own farms.

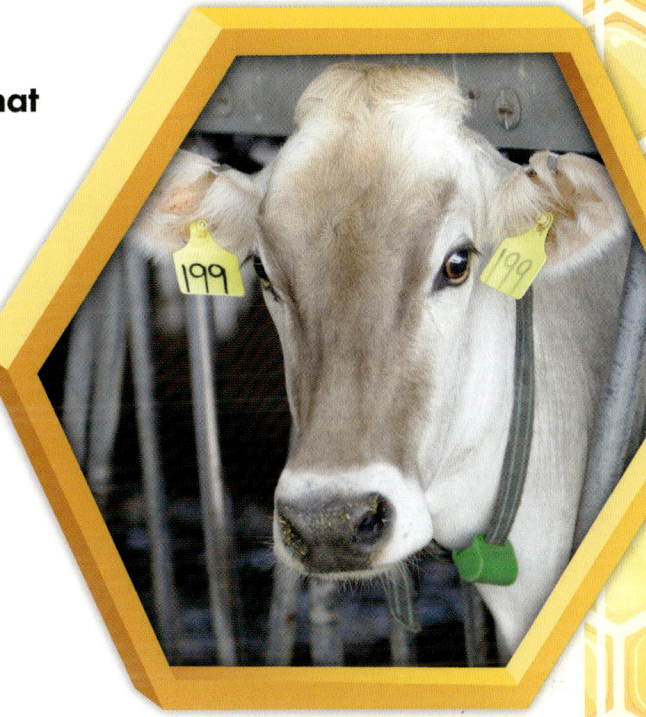

TIME FOR AN ICE CREAM BREAK! I'M THINKIN' COOKIES AND CREAM.

WHAT'S THE BUZZ?

Arizona ranchers produce enough beef for every person in our state to chow down on a burger each day for 300 days in a row! That's a pile of hamburgers!

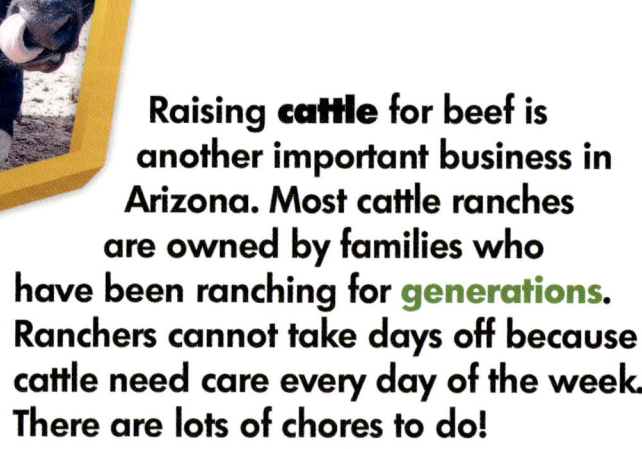

Raising **cattle** for beef is another important business in Arizona. Most cattle ranches are owned by families who have been ranching for **generations**. Ranchers cannot take days off because cattle need care every day of the week. There are lots of chores to do!

GET ALONG, LITTLE DOGIES!

©David W. Schafer, AGCA Photo Contest

When cattle have eaten most of the grass in one area, ranchers travel on horseback or in four-wheel drive vehicles to move the herd to another **pasture**. Moving the cattle gives them plenty of fresh grass to **graze**. The grass in the pasture they leave grows back, so the cattle can graze there again.

Ranchers also feed their cattle grain and hay, and they make sure their animals always have fresh drinking water.

©Brent Murphree

LET'S HELP SAVE ARIZONA'S WATER! TURN OFF THE FAUCET WHEN YOU BRUSH YOUR TEETH!

©Tim Trumble Photography

Farmers and ranchers in Arizona have always worried about water. Because there isn't much rain, they must find other ways to supply water to their crops and animals. Some farms and ranches get water from wells dug deep in the ground. Arizona also has many **dams** and **reservoirs** connected to **canal** systems that carry water to fields and pastures.

©Monica Kilcullen Pastor

Along with **conserving** water, it's also important to preserve our land! Arizona ranchers often set aside some of their property to protect **wildlife** and endangered species. This is a perfect example of how ranchers care about the environment.

©Dave Massey

Bees are really important to Arizona's amazing agricultural industry. Farmers depend on bees to pollinate their fields and orchards. When bees move **pollen** from one plant to another, it helps the plants produce crops. Bees also take pollen back to their beehives to eat.

Bees have other special jobs, too. They make honey and beeswax. Honey is a tasty, sweet treat that people love to eat. Beeswax is used for candles, crayons, soap, and furniture polish.

WHAT'S THE BUZZ?

Whenever you color with a crayon, polish a table, or dip your chicken strip in honey sauce, remember to thank a bee!

©Linda Covey

WHOA ... LOOK AT THE TIME!

I need to **Z-Z-Z-OOM** home so I won't be late for dinner. Now that we have traveled together, you can see why I love Arizona agriculture! What agricultural products do *you* love?

AGRICULTURE GLOSSARY

Acre: how land is measured; one acre is about the size of a football field

Canal: a man-made waterway for irrigation

Climate: the weather; temperature, rain, wind, snow, and other outside conditions

Conserve: to keep safe, save, or take care of

Conveyor Belt: a machine with a moving surface that takes products from place to place

Crop: plants that are grown and harvested for food

Dam: a barrier that holds back water

Dogie: a motherless calf in a herd

Fertilizer: a substance added to soil so plants will be well nourished

Generation: years in the life of a family; from babies to the time they become adults

Graze: to feed on plants throughout the day, like cattle and horses do

Harvest: gathering a crop when it is ready

Hay: dried stems and leaves of grass or alfalfa used to feed animals

Natural: not made by people

Nutritionist: a person who knows all about healthy foods

Pasture: land covered with plants that livestock eat

Pollen: a fine, yellow dust that plants make

Pollinate: the movement of pollen to and from plants by insects, animals, and wind

Reservoir: land where a large supply of water is collected

Straw: dried stems of wheat, barley, or oats used for animal bedding

Wildlife: animals and plants living in a natural area

CURRICULUM GUIDE
FOR TEACHERS, LIBRARIANS, AND PARENTS

READ ALOUD AND LOVE IT!

Prep Time

- Read and familiarize yourself with the book ahead of time.
- Preview the glossary.
- Prepare visual aids and supplementary activities.

Meet the Reader — Meet Pee Wee Bee

- Introduce yourself. Explain why you are the perfect person to share this book with your audience.
- Introduce the book, its title, the author, and the illustrator. Use and define these words to help children develop a better understanding of literary terminology.
- Before you start to read, ask the audience to predict what this book's topic might be.

Read-Aloud Tips

- Avoid the most common presentation mistake—reading too fast. Slow down and allow time for listeners to comprehend the information being shared.
- Pay close attention to punctuation and demonstrate how it impacts the story. Pause at commas. Stop at periods. Make your voice rise at question marks. Emphasize exclamation points.
- Maintain eye contact with the children as you share the story. Make your eyes expressive.

Ham It Up

- Vary your voice. Change your volume, tempo, and inflection (i.e. whisper, laugh, emphasize key words, convey enthusiasm and excitement).
- Use dynamic body language to engage the audience—wave, point, flap your arms, nod your head, etc.
- Use a special, animated voice whenever Pee Wee Bee talks.
- Be dramatic. Don't be afraid to be a little bit silly. Children love it!

Get Engaged

- "Echo read" by inviting listeners to repeat Bee's comments, key terms, or fun facts.
- Have the kids say *"BZZZZZ"* whenever you turn a page.
- Let children take turns holding the book/turning the pages.
- Bring products (cotton bolls, eggs, pecans, lettuce, etc.) or photos to share as visual aids. Select "helpers" to stand and display the products/photos when they are introduced in the book.
- Ask open-ended questions throughout the book:
    - "Why does Pee Wee Bee think she is so important to this crop?"
    - "How does technology help the agricultural industry?"
    - "What do you think would happen if there were no farmers?"

Comprehension Questions

Based on the lesson time allocated and the age of your audience, select appropriate extension topics from the list below. You can present them while you are reading and/or when you finish the book.

- Compare beef cattle and dairy cows.
- Cotton farmers work hard. What are some of the chores you think they must do every day?
- Name three specialty crops.
- Is it correct to say "cotton boll" or "cotton ball"?
- Describe some of the technologies that ranchers and farmers use today.
- How many jobs can you name that are tied to Arizona agriculture?
- Name as many dairy products as you can.
- Which Arizona agriculture job would you like to do? Why?
- How can you help save Arizona's water?
- If you had to write a report about Arizona agriculture, which topics would you choose?

Online Resources

Would you like a supplemental activity that ties to Pee Wee Bee? A lesson plan? A song? A recipe? Check out these helpful links:

- Water-related resources: http://arizonawet.arizona.edu
- Central Arizona Project (water systems): http://www.cap-az.com
- University of Arizona Cooperative Extension: http://cals.arizona.edu/agliteracy
- Arizona Farm Bureau: http://www.azfb.org
- "Ag in the Classroom" lesson plans and resources for grades Pre-K to 12: http://www.agclassroom.org

ABOUT THE CO-AUTHORS

Bonnie Apperson Jacobs

Bonnie Apperson Jacobs graduated from Arizona State University with bachelor's and master's degrees in education and served as a classroom teacher, media specialist, and school principal for 41 years. Her enthusiastic, maverick style of narrating books for children has been Bonnie's professional trademark. Her skills as a journalist and writer have been widely recognized.

Terri Mainwaring

A lifelong educator, Terri Mainwaring holds undergraduate and master's degrees from Arizona State University and has widespread experience both as a classroom teacher and as a school and district-level administrator. Her hobbies and interests include traveling with her husband, Tom, enjoying her grandchildren, Brady and Aly, reading, writing, and exercising with friends at the YMCA.

ABOUT THE COOPERATIVE PARTNERS

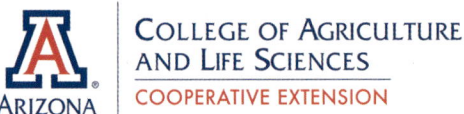

ARIZONA | COLLEGE OF AGRICULTURE AND LIFE SCIENCES
COOPERATIVE EXTENSION

The University of Arizona College of Agriculture and Life Sciences Cooperative Extension (CALS-CE) serves Arizona's fifteen counties and five tribal reservations, providing integral educational programs for all Arizonans. Tasked with "Improving Lives and Communities," CALS-CE is a part of a nationwide educational network of scientists and educators who help people solve problems and put knowledge to use, improving lives, families, communities, the environment, and economies in Arizona and beyond.

Maricopa County Farm Bureau

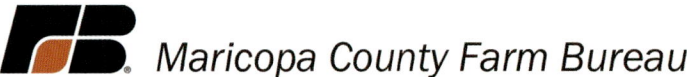

The Maricopa County Farm Bureau (MCFB) is one of thirteen county farm bureaus within the Arizona Farm Bureau Federation. A grassroots effort dedicated to promoting and protecting agriculture, MCFB is a nonprofit organization governed by an elected board of directors who are local agriculture producers. MCFB concentrates on four core areas: representing agriculture in legislative and regulatory actions; education and communication; member benefits; and developing leaders for the agriculture industry.

CALS-CE/MCFB Partnership

The University of Arizona College of Agriculture and Life Sciences Cooperative Extension (UA CALS-CE) and the Maricopa County Farm Bureau (MCFB) collaborate to teach children about the sources of their food, clothes, and plants that beautify their homes. A MCFB gift provided initial funds for this book. Jeannette Fish, executive director of MCFB, and Monica Kilcullen Pastor, associate programmatic agent, ag and natural resources of CALS, provided technical expertise to the authors. Additional financial support for the book was provided by individuals and organizations that support educating children about Arizona agriculture.

ABOUT THE PUBLISHER

Helping authors of children's books shape the future with great reads for growing minds...

Linda F. Radke
President

Shining brightly since 1985, Five Star Publications, Inc. is proud of its reputation for excellence in producing and marketing award-winning books for adults and children worldwide. The genres represented in its growing collection include educational titles, nonfiction, picture books, juvenile fiction, memoirs, Westerns, novels, professional "how-to's," and more.

Having assembled a team of dozens of skilled industry professionals, Linda F. Radke, founder and president of Five Star Publications, Inc., is committed to helping both established and aspiring authors of all ages continually reach new heights. Along with providing book production/marketing services, the Five Star team also assists organizations with website redesign, logo design, and corporate/product branding.

Setting the bar for partnership publishing and professionally fulfilling traditional publishing contracts, Five Star Publications is recognized as an industry leader in creativity, innovation, and customer service.

Many Five Star Publications titles have been recognized on local, national, and international levels, and their authors have enjoyed engaging in promotional opportunities in schools, corporations, and media venues across America.

A division of Five Star Publications, Inc., Little Five Star is a proud publisher of great books for growing minds and excels in educating and entertaining young thinkers. Little Five Star helps children make better choices and accept themselves and others.

 **www.LittleFiveStar.com**

ORDER FORM
Add these titles to your Little Five Star collection!

ITEM	Retail Price	QTY	TOTAL
Arizona Agriculture: Bee's Amazing Adventure by Bonnie Apperson Jacobs and Terri Mainwaring (ISBN: 978-1-58985-267-9)	$11.95 US $12.95 CAN		
Arizona Way Out West & Wacky: Awesome Activities, Humorous History and Fun Facts! by Lynda Exley and Conrad J. Storad (ISBN: 978-1-58985-047-7)	$11.95 US $12.95 CAN		
Burton the Scarecrow–Friendship Tales from the Farm Series (Books 1-5) by V.A. Boeholt	$59.75 US $66.75 CAN		
Cheery: The True Adventures of a Chiricahua Leopard Frog by Elizabeth W. Davidson, Ph.D. (ISBN: 978-1-58985-025-5)	$11.95 US $12.95 CAN		
Gator, Gator, Second Grader (Classroom Pet...or Not?) by Conrad J. Storad (ISBN: 978-1-58985-271-6)	$9.95 US $11.95 CAN		
GQ GQ. Where Are You? Adventures of a Gambel's Quail by Sharon I. Ritt (ISBN: 978-1-58985-223-5)	$14.95 US $15.95 CAN		
The Moon Saw It All by Nancy L. Young (ISBN: 978-1-58985-250-1)	$11.95 US $12.95 CAN		
Rattlesnake Rules by Conrad J. Storad (ISBN: 978-1-58985-211-2)	$7.95 US/CAN		
	Subtotal		
* 8.8% sales tax - on all orders originating in Arizona.	*Tax		
Ground shipping: *$8.00 or 10% of the order (whichever is greater) Allow 1 to 2 weeks for delivery.			
Mail form to: Five Star Publications, Inc. P.O. Box 6698, Chandler, AZ 85246-6698	TOTAL		

Name:

Address:

City, State, Zip:

Daytime Phone: Fax:

Email:

Method of Payment: ☐ VISA ☐ MasterCard ☐ Discover Card ☐ American Express

Account Number: Expiration Date:

Signature: 3 or 4 Digit Security Number:

Great Books for Growing Minds
A Division of Five Star Publications, Inc.
P.O. Box 6698 • Chandler, AZ 85246-6698
(480) 940-8182 or (866) 471-0777
Fax: (480) 940-8787
info@FiveStarPublications.com
www.LittleFiveStar.com

For information on how to order an ebook go to:
 eStarPublish.com

How were you referred to Five Star Publications, Inc.?
☐ Friend ☐ Internet ☐ Book Event ☐ Other

Book an author for your event today!
Visit **www.SchoolBookings.com** *for more info.*